GRAVITY

Susan Canizares • Daniel Moreton

Scholastic Inc.
New York • Toronto • London • Auckland • Sydney

Acknowledgments

Literacy Specialist: Linda Cornwell

National Science Consultant: David Larwa

Design: Silver Editions

Photo Research: Amla Sanghvi

Endnotes: Cleo Cacoulidis

Endnote Illustrations: Craig Spearing

───────────────────

Photographs: Cover: Tom Mareschal/Image Bank; p. 1: Frank Orel/Tony Stone Images; p. 2: Andy Roberts/Tony Stone Images; p. 3: Gary Holscher/Tony Stone Images; p. 4: H. Richard Johnston/FPG International; p. 5: Paul Dance/Tony Stone Images; pp. 6–7: Roy Morsch/The Stock Market; p. 8: Marc Muench/Tony Stone Images; p. 9: G. Brad Lewis/Tony Stone Images; p. 10: Laurie Rubin/Image Bank; p. 11: Tony Freeman/Photo Edit; p. 12: D. Young-Wolff/Photo Edit.

Library of Congress Cataloging-in-Publication Data
Canizares, Susan 1960-
Gravity/Susan Canizares, Daniel Moreton.
p.cm. --(Science emergent readers)
Summary: Simple text and photographs present the effects of gravity in making things come down, from falling seeds to dripping wax.
ISBN 0-439-08127-0 (pbk.: alk. paper)
1. Gravitation--Juvenile literature. 2. Gravity--Juvenile literature. [1. Gravity.] I. Moreton, Daniel. II. Title. III. Series.
QC178.C175 1999
531'.14--dc21
98-53318
CIP AC

13 14 15 16 17 18 19 20   08   08 07 06 05 04

Things come down.

Seeds blow down.

Icicles hang down.

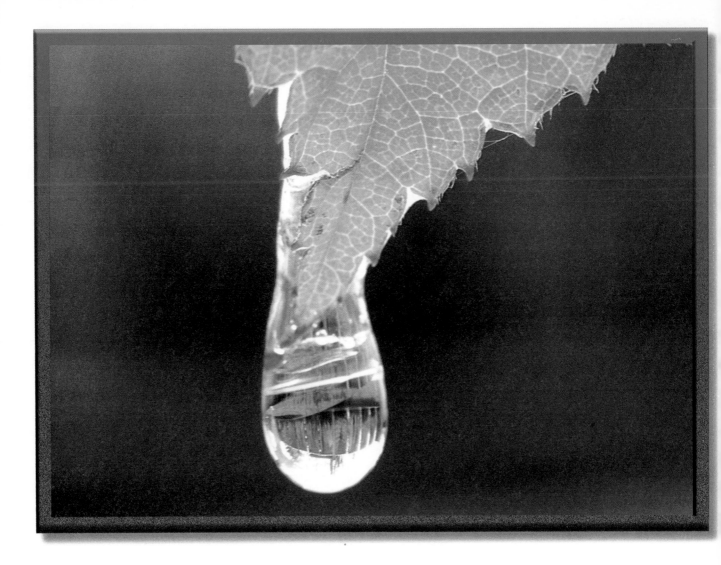

Water drops down.

Tears stream down.

Rain pours down.

Snow slides down.

Lava flows down.

Wax drips down.

Leaves fall down.

Gravity makes things come down.

GRAVITY

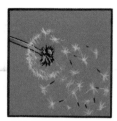

Things come down When you throw a ball up in the air, why does it fall back down? No matter how hard or far you throw the ball, it always drops back to the ground. The ball falls because of an invisible force called gravity. Gravity is everywhere, yet we can't see, touch, smell, or hear it. But we know gravity surrounds us because we can see it work. When an apple falls from a tree or you hold up a pencil and let it go, it is gravity that pulls the object down and keeps it on the ground. Without gravity, people, animals, plants, buildings, and even oceans would float away into space.

Seeds blow down The wind can blow seeds for miles and miles through the air. But no matter how far the seeds travel, gravity will eventually pull them back down. Once seeds fall to the ground, they grow roots in the soil, and in a few months they'll grow into new plants. In the spring, you'll see the new plants bloom in the places where the seeds fell.

Icicles hang down In wintertime, when it's very cold, raindrops freeze as they drip down from surfaces they land on. These frozen drops form icicles. When spring comes and it is warm again, the icicles melt and fall to the ground. Heat from the sun causes water on the ground to evaporate. The water rises through the air, gathers together, and forms clouds. When the clouds are full, the water falls in drops back to the earth as rain or snow and the cycle starts all over again.

Tears stream down Tears are drops of clear, salty liquid that fall from our eyes. They help keep our eyes wet and clean and free from bacteria. Tears flow when we are sad, and sometimes even when we are happy.

Rain pours down Plants need water to grow and be healthy. If they don't get enough water from rain, they may wilt and die. Flowers and trees absorb water through their roots.

When the rain pours down from the clouds, it showers the earth's plants with the water they need to survive. Whether it is a light drizzle or a heavy rainstorm, rain always falls down.

Snow slides down Some mountains are extremely tall and their peaks reach far into the sky. At the top of these big mountains, winter snow stays on the ground all year long because the air is always cold. If the snow piles up too high or if something else disturbs it, gravity causes the snow to slide down. When a lot of snow crashes down the mountainside, it is called an avalanche.

Lava flows down Deep inside the earth it is so hot that molten rocks become a fiery substance called lava. Lava comes out of a volcano or a crack in the earth's surface. The burning lava can shoot up out of a volcano during an eruption, but gravity always makes it come flowing down again.

Wax drips down The first candles were made more than 2,000 years ago. People used them to light their homes. A candle is made from a piece of string, called a wick, that is covered in wax. The heat from a burning candle melts the wax, which drips down. Today we like to use candles on birthday cakes. But you have to make sure to blow them out quickly; otherwise gravity will pull the melting wax down onto your frosting!

Leaves fall down During winter there are fewer hours of light each day than in spring or summer. It is colder, too. The shorter hours and colder weather tell certain trees that it is time to take a rest. The trees do this by dropping their leaves. The force of gravity pulls the dead leaves to the ground, where they fertilize the soil. In spring, new green leaves will sprout on the branches again.

Gravity makes things come down The invisible force of gravity brings everything down and keeps people and objects from drifting into space. Gravity is all around us. And we know it's working when we see raindrops fall from the sky and make puddles on the ground that we can splash in with our friends. Without gravity, where would we be?